I0481385

 equals(me)

Radicals Practice
1st Edition

equals(me) packets give exciting and challenging entertainment while improving ones mind skills.

equals(me) packets are designed to assist in learning math and algebra concepts. These packets provide excellent algebra practice.

Each packet includes a variety of math and algebra techniques. The packets are modeled using the Study-Try-Compare methods. This gives you the opportunity to study the steps in solving, try for yourself, and compare results. You can also try to solve and then compare your solution.

For school, college, work, or at home, equals(me) packets make math fun.

Disclaimer: The author makes no guarantees as to the accuracy of the material included. This packet is for entertainment purposes only. The included solutions are only that of which the author would perform. There may be several solution methods and/or answers to the equations and/or expressions included in this packet. .

ISBN: 978-1984041722

0118rev 1

Simplify Radicals

Type 1 Radical

$\sqrt{32}$

Perfect Squares

Find a perfect square and another number whose product equals the radicand.

$\sqrt{16} \cdot \sqrt{2}$

$\sqrt{4} = 2$
$\sqrt{9} = 3$
$\sqrt{16} = 4$

Continue to factor.

$\sqrt{4} \cdot \sqrt{4} \cdot \sqrt{2}$

$\sqrt{25} = 5$
$\sqrt{36} = 6$

$2 \cdot 2 \cdot \sqrt{2}$

$\sqrt{49} = 7$
$\sqrt{64} = 8$

$4\sqrt{2}$

$\sqrt{81} = 9$

Type 2 Radical

$\frac{4}{\sqrt{2}}$

Eliminate radical denominator by multiplying numerator and denominator by the denominator.

$\frac{4}{\sqrt{2}} \cdot \frac{\sqrt{2}}{\sqrt{2}} = \frac{4\sqrt{2}}{\sqrt{4}} = \frac{4\sqrt{2}}{2}$

Simplify numerator and denominator. $\frac{4\sqrt{2}}{2} = 2\sqrt{2}$

Type 3 Radical (Rationalize)

$\frac{4 + \sqrt{3}}{5 - \sqrt{3}}$

Multiply the numerator and denominator by the conjugate of denominator.

$\frac{4 + \sqrt{3}}{5 - \sqrt{3}} \cdot \frac{5 + \sqrt{3}}{5 + \sqrt{3}} = \frac{23 + 9\sqrt{3}}{22}$

Solve Radical Equations

Solving for x.

$$\sqrt{4x + 8} = 6$$

Square both sides to eliminate radical.

$$\sqrt{(4x + 8)^2} = 6^2$$

$$4x + 8 = 36$$

Combine like terms.

$$4x = 36 - 8$$
$$4x = 28$$
$$x = \frac{28}{4}$$

Factoring

Find roots of expression

$$x^2 + 5x + 6 = (x + 2)(x + 3)$$
$$using \ x^2 + (m+n)x + mn$$

Use FOIL to check:

$$(x + 2)(x + 3)$$
$$x^2 + 2x + 3x + 6$$

Combining like terms:

$$x^2 + 5x + 6$$

See pages 80 & 81 for more rules of algebra.

Simplify Radical Expressions

$$\sqrt{32}$$
$$= \sqrt{4} \cdot \sqrt{8}$$
$$= 2 \cdot \sqrt{4} \cdot \sqrt{2}$$
$$= 2 \cdot 2 \cdot \sqrt{2}$$
$$= 4\sqrt{2}$$

$$\sqrt{125}$$
$$= \sqrt{25} \cdot \sqrt{5}$$
$$= 5 \cdot \sqrt{5}$$
$$= 5\sqrt{5}$$

$$\sqrt{180}$$
$$= \sqrt{4} \cdot \sqrt{45}$$
$$= 2 \cdot \sqrt{9} \cdot \sqrt{5}$$
$$= 2 \cdot 3 \cdot \sqrt{5}$$
$$= 6\sqrt{5}$$

$$\sqrt{240}$$
$$= \sqrt{4} \cdot \sqrt{60}$$
$$= 2 \cdot \sqrt{4} \cdot \sqrt{15}$$
$$= 2 \cdot 2 \cdot \sqrt{15}$$
$$= 4\sqrt{15}$$

Simplify Radical Expressions

$$\sqrt{216}$$

$$\sqrt{512}$$

$$5\sqrt{54}$$

$$4\sqrt{80}$$

Simplify Radical Expressions

$3\sqrt{90}$
$= 3 \cdot \sqrt{9} \cdot \sqrt{10}$
$= 3 \cdot 3 \cdot \sqrt{10}$
$= 9\sqrt{10}$

$5\sqrt{147}$
$= 5 \cdot \sqrt{49} \cdot \sqrt{3}$
$= 5 \cdot 7 \cdot \sqrt{3}$
$= 35\sqrt{3}$

$4\sqrt{72}$
$= 4 \cdot \sqrt{36} \cdot \sqrt{2}$
$= 4 \cdot 6 \cdot \sqrt{2}$
$= 24\sqrt{2}$

$3\sqrt{400}$
$= 3 \cdot \sqrt{25} \cdot \sqrt{16}$
$= 3 \cdot 5 \cdot 4$
$= 60$

Simplify Radical Expressions

$$3\sqrt{90}$$

_____

$$5\sqrt{147}$$

_____

$$4\sqrt{72}$$

_____

$$3\sqrt{400}$$

_____

Simplify Radical Expressions

$$\sqrt{512y^2}$$
$$= y \cdot \sqrt{16} \cdot \sqrt{32}$$
$$= y \cdot 4 \cdot \sqrt{16} \cdot \sqrt{2}$$
$$= y \cdot 4 \cdot 4 \cdot \sqrt{2}$$
$$= 16y\sqrt{2}$$

$$\sqrt{512k^3}$$
$$= k \cdot \sqrt{16} \cdot \sqrt{32k}$$
$$= k \cdot 4 \cdot \sqrt{16} \cdot \sqrt{2k}$$
$$= k \cdot 4 \cdot 4 \cdot \sqrt{2k}$$
$$= 16k\sqrt{2k}$$

$$\sqrt{216y^4}$$
$$= y^2 \cdot \sqrt{36} \cdot \sqrt{6}$$
$$= y^2 \cdot 6 \cdot \sqrt{6}$$
$$= 6y^2\sqrt{6}$$

$$6\sqrt{72y^2}$$
$$= 6 \cdot y \cdot \sqrt{36} \cdot \sqrt{2}$$
$$= 6 \cdot y \cdot 6 \cdot \sqrt{2}$$
$$= 36y\sqrt{2}$$

Simplify Radical Expressions

$$\sqrt{32}$$

$$\sqrt{125}$$

$$\sqrt{180}$$

$$\sqrt{240}$$

Simplify Radical Expressions

$$\sqrt{360}$$
$$= \sqrt{4} \cdot \sqrt{90}$$
$$= 2 \cdot \sqrt{9} \cdot \sqrt{10}$$
$$= 2 \cdot 3 \cdot \sqrt{10}$$
$$= 6\sqrt{10}$$

$$\sqrt{425}$$
$$= \sqrt{25} \cdot \sqrt{17}$$
$$= 5 \cdot \sqrt{17}$$
$$= 5\sqrt{17}$$

$$\sqrt{140}$$
$$= \sqrt{4} \cdot \sqrt{35}$$
$$= 2 \cdot \sqrt{35}$$
$$= 2\sqrt{35}$$

$$\sqrt{160}$$
$$= \sqrt{16} \cdot \sqrt{10}$$
$$= \sqrt{4} \cdot \sqrt{4} \cdot \sqrt{10}$$
$$= 2 \cdot 2 \cdot \sqrt{10}$$
$$= 4\sqrt{10}$$

Simplify Radical Expressions

$$\sqrt{360}$$

_____
_____
_____
_____
_____

$$\sqrt{425}$$

_____
_____
_____
_____
_____

$$\sqrt{140}$$

_____
_____
_____
_____
_____

$$\sqrt{160}$$

_____
_____
_____
_____
_____

Simplify Radical Expressions

$\sqrt{216}$
$= \sqrt{36} \cdot \sqrt{6}$
$= 6 \cdot \sqrt{6}$
$= 6\sqrt{6}$

$\sqrt{512}$
$= \sqrt{16} \cdot \sqrt{32}$
$= 4 \cdot \sqrt{4} \cdot \sqrt{8}$
$= 4 \cdot 2 \cdot \sqrt{4} \cdot \sqrt{2}$
$= 4 \cdot 2 \cdot 2 \cdot \sqrt{2}$
$= 16\sqrt{2}$

$5\sqrt{54}$
$= 5 \cdot \sqrt{9} \cdot \sqrt{6}$
$= 5 \cdot 3 \cdot \sqrt{6}$
$= 15\sqrt{6}$

$4\sqrt{80}$
$= 4 \cdot \sqrt{4} \cdot \sqrt{20}$
$= 4 \cdot 2 \cdot \sqrt{4} \cdot \sqrt{5}$
$= 4 \cdot 2 \cdot 2 \cdot \sqrt{5}$
$= 16\sqrt{5}$

Simplify Radical Expressions

$$\sqrt{216}$$

_____
_____
_____
_____
_____

$$\sqrt{512}$$

_____
_____
_____
_____
_____

$$5\sqrt{54}$$

_____
_____
_____
_____
_____

$$4\sqrt{80}$$

_____
_____
_____
_____
_____

Simplify Radical Expressions

$3\sqrt{90}$

$= 3 \cdot \sqrt{9} \cdot \sqrt{10}$

$= 3 \cdot 3 \cdot \sqrt{10}$

$= 9\sqrt{10}$

$5\sqrt{147}$

$= 5 \cdot \sqrt{49} \cdot \sqrt{3}$

$= 5 \cdot 7 \cdot \sqrt{3}$

$= 35\sqrt{3}$

$4\sqrt{72}$

$= 4 \cdot \sqrt{36} \cdot \sqrt{2}$

$= 4 \cdot 6 \cdot \sqrt{2}$

$= 24\sqrt{2}$

$3\sqrt{400}$

$= 3 \cdot \sqrt{25} \cdot \sqrt{16}$

$= 3 \cdot 5 \cdot 4$

$= 60$

Simplify Radical Expressions

$$3\sqrt{90}$$

$$5\sqrt{147}$$

$$4\sqrt{72}$$

$$3\sqrt{400}$$

Simplify Radical Expressions

$$\sqrt{512y^2}$$
$$= y \cdot \sqrt{16} \cdot \sqrt{32}$$
$$= y \cdot 4 \cdot \sqrt{16} \cdot \sqrt{2}$$
$$= y \cdot 4 \cdot 4 \cdot \sqrt{2}$$
$$= 16y\sqrt{2}$$

$$\sqrt{512k^3}$$
$$= k \cdot \sqrt{16} \cdot \sqrt{32k}$$
$$= k \cdot 4 \cdot \sqrt{16} \cdot \sqrt{2k}$$
$$= k \cdot 4 \cdot 4 \cdot \sqrt{2k}$$
$$= 16k\sqrt{2k}$$

$$\sqrt{216y^4}$$
$$= y^2 \cdot \sqrt{36} \cdot \sqrt{6}$$
$$= y^2 \cdot 6 \cdot \sqrt{6}$$
$$= 6y^2\sqrt{6}$$

$$6\sqrt{72y^2}$$
$$= 6 \cdot y \cdot \sqrt{36} \cdot \sqrt{2}$$
$$= 6 \cdot y \cdot 6 \cdot \sqrt{2}$$
$$= 36y\sqrt{2}$$

Simplify Radical Expressions

$$\sqrt{512y^2}$$

_____
_____
_____
_____
_____

$$\sqrt{512k^3}$$

_____
_____
_____
_____
_____

$$\sqrt{216y^4}$$

_____
_____
_____
_____

$$6\sqrt{72y^2}$$

_____
_____
_____
_____

Simplify Radical Expressions

$$\sqrt{63x^3}$$
$$= x \cdot \sqrt{9} \cdot \sqrt{7x}$$
$$= x \cdot 3 \cdot \sqrt{7x}$$
$$= 3x\sqrt{7x}$$

$$3\sqrt{36b^2}$$
$$= 3b \cdot \sqrt{4} \cdot \sqrt{9}$$
$$= 3b \cdot 2 \cdot 3$$
$$= 3b \cdot 6$$
$$= 18b$$

$$x\sqrt{81x^2}$$
$$= x^2 \cdot \sqrt{9} \cdot \sqrt{9}$$
$$= x^2 \cdot 3 \cdot 3$$
$$= 9x^2$$

$$y\sqrt{108b^2}$$
$$= y \cdot b \cdot \sqrt{9} \cdot \sqrt{12}$$
$$= y \cdot b \cdot 3 \cdot \sqrt{4} \cdot \sqrt{3}$$
$$= y \cdot b \cdot 3 \cdot 2 \cdot \sqrt{3}$$
$$= 6yb\sqrt{3}$$

Simplify Radical Expressions

$$\sqrt{63x^3}$$

_____
_____
_____
_____
_____

$$3\sqrt{36b^2}$$

_____
_____
_____
_____
_____

$$x\sqrt{81x^2}$$

_____
_____
_____
_____
_____

$$y\sqrt{108b^2}$$

_____
_____
_____
_____
_____

Simplify Radical Expressions

$$x \sqrt{50x^2}$$
$$= x \cdot x \cdot \sqrt{25} \cdot \sqrt{2}$$
$$= x \cdot x \cdot 5 \cdot \sqrt{2}$$
$$= 5x^2\sqrt{2}$$

$$2x \sqrt{63x^2}$$
$$= 2x \cdot x \cdot \sqrt{9} \cdot \sqrt{7}$$
$$= 2x \cdot x \cdot 3 \cdot \sqrt{7}$$
$$= 2x^2 \cdot 3 \cdot \sqrt{7}$$
$$= 6x^2\sqrt{7}$$

$$3y \sqrt{64y^4}$$
$$= 3y \cdot y^2 \cdot \sqrt{16} \cdot \sqrt{4}$$
$$= 3y \cdot y^2 \cdot 4 \cdot 2$$
$$= 3y^3 \cdot 8$$

$$5x \sqrt{150x^3}$$
$$= 5x \cdot x \cdot \sqrt{25} \cdot \sqrt{6x}$$
$$= 5x \cdot x \cdot 5 \cdot \sqrt{6x}$$
$$= 5x^2 \cdot 5 \cdot \sqrt{6x}$$
$$= 25x^2 \sqrt{6x}$$

Simplify Radical Expressions

$$x \sqrt{50x^2}$$

_____

_____

_____

_____

_____

_____

$$2x \sqrt{63x^2}$$

_____

_____

_____

_____

_____

$$3y \sqrt{64y^4}$$

_____

_____

_____

_____

_____

$$5x \sqrt{150x^3}$$

_____

_____

_____

_____

_____

Simplify Radical Expressions

$$\sqrt{8} \cdot (\sqrt{8} + 5)$$
$$= \sqrt{8^2} + 5\sqrt{8}$$
$$= \sqrt{64} + 5\sqrt{8}$$
$$= 8 + 5 \cdot \sqrt{4} \cdot \sqrt{2}$$
$$= 8 + 5 \cdot 2 \cdot \sqrt{2}$$
$$= 8 + 10\sqrt{2}$$

$$\sqrt{2} \cdot (\sqrt{8} - 7)$$
$$= \sqrt{16} - 7\sqrt{2}$$
$$= 4 - 7\sqrt{2}$$

$$\sqrt{12} \cdot (\sqrt{12} - 2)$$
$$= \sqrt{12^2} - 2\sqrt{12}$$
$$= \sqrt{144} - 2\sqrt{12}$$
$$= 12 - 2 \cdot \sqrt{4} \cdot \sqrt{3}$$
$$= 12 - 2 \cdot 2 \cdot \sqrt{3}$$
$$= 12 - 4\sqrt{3}$$

$$\sqrt{16} \cdot (\sqrt{16} - 12)$$
$$= \sqrt{16^2} - 12\sqrt{16}$$
$$= \sqrt{256} - 12\sqrt{16}$$
$$= 16 - 12 \cdot 4$$
$$= -32$$

Simplify Radical Expressions

$$\sqrt{8} \cdot (\sqrt{8} + 5)$$

$$\sqrt{2} \cdot (\sqrt{8} - 7)$$

$$\sqrt{12} \cdot (\sqrt{12} - 2)$$

$$\sqrt{16} \cdot (\sqrt{16} - 12)$$

Simplify Radical Expressions

$$\sqrt{2} \cdot (\sqrt{2} + 4)$$
$$= \sqrt{2^2} + 4\sqrt{2}$$
$$= \sqrt{4} + 4\sqrt{2}$$
$$= 2 + 4\sqrt{2}$$

$$\sqrt{3} \cdot (\sqrt{3} + 8)$$
$$= \sqrt{3^2} + 8\sqrt{3}$$
$$= \sqrt{9} + 8\sqrt{3}$$
$$= 3 + 8\sqrt{3}$$

$$\sqrt{5} \cdot (\sqrt{5} - 4)$$
$$= \sqrt{5^2} - 4\sqrt{5}$$
$$= \sqrt{25} - 4\sqrt{5}$$
$$= 5 - 4\sqrt{5}$$

$$\sqrt{7} \cdot (\sqrt{7} - 9)$$
$$= \sqrt{7^2} - 9\sqrt{7}$$
$$= \sqrt{49} - 9\sqrt{7}$$
$$= 7 - 9\sqrt{7}$$

Simplify Radical Expressions

$$\sqrt{2} \cdot (\sqrt{2} + 4)$$

_____

$$\sqrt{3} \cdot (\sqrt{3} + 8)$$

_____

$$\sqrt{5} \cdot (\sqrt{5} - 4)$$

_____

$$\sqrt{7} \cdot (\sqrt{7} - 9)$$

_____

Simplify Radical Expressions

$$(\sqrt{2} + 3)(\sqrt{2} + 5)$$
$$= \sqrt{2^2} + 5\sqrt{2} + 3\sqrt{2} + 15$$
$$= \sqrt{4} + 5\sqrt{2} + 3\sqrt{2} + 15$$
$$= 2 + 5\sqrt{2} + 3\sqrt{2} + 15$$
$$= 17 + 8\sqrt{2}$$

$$(\sqrt{5} - 2)(\sqrt{5} + 2)$$
$$= \sqrt{5^2} + 2\sqrt{5} - 2\sqrt{5} - 4$$
$$= \sqrt{25} + 2\sqrt{5} - 2\sqrt{5} - 4$$
$$= 5 + 2\sqrt{5} - 2\sqrt{5} - 4$$
$$= 1$$

$$(7 - \sqrt{2})(\sqrt{2} + 3)$$
$$= 7\sqrt{2} + 21 - \sqrt{2^2} - 3\sqrt{2}$$
$$= 7\sqrt{2} + 21 - \sqrt{4} - 3\sqrt{2}$$
$$= 7\sqrt{2} + 21 - 2 - 3\sqrt{2}$$
$$= 19 + 4\sqrt{2}$$

$$(2\sqrt{2} + 1)(\sqrt{2} + 5)$$
$$= 2\sqrt{2^2} + 10\sqrt{2} + \sqrt{2} + 5$$
$$= 2\sqrt{4} + 10\sqrt{2} + \sqrt{2} + 5$$
$$= 4 + 10\sqrt{2} + \sqrt{2} + 5$$
$$= 9 + 11\sqrt{2}$$

Simplify Radical Expressions

$$(\sqrt{2} + 3)(\sqrt{2} + 5)$$

_____

_____

_____

_____

$$(\sqrt{5} - 2)(\sqrt{5} + 2)$$

_____

_____

_____

_____

$$(7 - \sqrt{2})(\sqrt{2} + 3)$$

_____

_____

_____

_____

$$(2\sqrt{2} + 1)(\sqrt{2} + 5)$$

_____

_____

_____

_____

Simplify Radical Expressions

$$(-\sqrt{3} + 2)(\sqrt{3} - 4)$$
$$= -\sqrt{3^2} + 4\sqrt{3} + 2\sqrt{3} - 8$$
$$= -\sqrt{9} + 4\sqrt{3} + 2\sqrt{3} - 8$$
$$= -3 + 4\sqrt{3} + 2\sqrt{3} - 8$$
$$= -11 + 6\sqrt{3}$$

$$(\sqrt{5} - 8)(-\sqrt{5} - 2)$$
$$= -\sqrt{5^2} - 2\sqrt{5} + 8\sqrt{5} + 16$$
$$= -\sqrt{25} - 2\sqrt{5} + 8\sqrt{5} + 16$$
$$= -5 - 2\sqrt{5} + 8\sqrt{5} + 16$$
$$= 11 + 6\sqrt{5}$$

$$(-\sqrt{7} + 1)(-\sqrt{7} + 9)$$
$$= \sqrt{7^2} - 9\sqrt{7} - \sqrt{7} + 9$$
$$= \sqrt{49} - 9\sqrt{7} - \sqrt{7} + 9$$
$$= 7 - 9\sqrt{7} - \sqrt{7} + 9$$
$$= 16 - 10\sqrt{7}$$

$$(3\sqrt{2} - 1)(4\sqrt{2} - 6)$$
$$= 12\sqrt{2^2} - 18\sqrt{2} - 4\sqrt{2} + 6$$
$$= 12\sqrt{4} - 18\sqrt{2} - 4\sqrt{2} + 6$$
$$= 24 - 18\sqrt{2} - 4\sqrt{2} + 6$$
$$= 30 - 22\sqrt{2}$$

Simplify Radical Expressions

$$(-\sqrt{3} + 2)(\sqrt{3} - 4)$$

$$(\sqrt{5} - 8)(-\sqrt{5} - 2)$$

$$(-\sqrt{7} + 1)(-\sqrt{7} + 9)$$

$$(3\sqrt{2} - 1)(4\sqrt{2} - 6)$$

Simplify Radical Expressions

$$(\sqrt{6} + \sqrt{3})(\sqrt{6} - \sqrt{3}) + 4$$
$$= \sqrt{6^2} - \sqrt{18} + \sqrt{18} - \sqrt{3^2} + 4$$
$$= \sqrt{36} - \sqrt{9} + 4$$
$$= 6 - 3 + 4$$
$$= 7$$

$$(\sqrt{5} - \sqrt{7})(\sqrt{5} + \sqrt{7}) + 8$$
$$= \sqrt{5^2} + \sqrt{35} - \sqrt{35} - \sqrt{7^2} + 8$$
$$= \sqrt{25} - \sqrt{49} + 8$$
$$= 5 - 7 + 8$$
$$= 6$$

$$(\sqrt{8} - \sqrt{2})(\sqrt{8} + \sqrt{2}) - 21$$
$$= \sqrt{8^2} + \sqrt{16} - \sqrt{16} - \sqrt{2^2} - 21$$
$$= \sqrt{64} - \sqrt{4} - 21$$
$$= 8 - 2 - 21$$
$$= -15$$

$$(2\sqrt{2} + \sqrt{3})(2\sqrt{2} - \sqrt{3}) + 12$$
$$= 4\sqrt{2^2} - 2\sqrt{6} + 2\sqrt{6} - \sqrt{3^2} + 12$$
$$= 4\sqrt{4} - \sqrt{9} + 12$$
$$= 8 - 3 + 12$$
$$= 17$$

Simplify Radical Expressions

$$(\sqrt{6} + \sqrt{3})(\sqrt{6} - \sqrt{3}) + 4$$

$$(\sqrt{5} - \sqrt{7})(\sqrt{5} + \sqrt{7}) + 8$$

$$(\sqrt{8} - \sqrt{2})(\sqrt{8} + \sqrt{2}) - 21$$

$$(2\sqrt{2} + \sqrt{3})(2\sqrt{2} - \sqrt{3}) + 12$$

Simplify Radical Expressions

$$(-2\sqrt{7} + \sqrt{5})(2\sqrt{7} + \sqrt{5})$$
$$= -4\sqrt{7^2} - 2\sqrt{35} + 2\sqrt{35} + \sqrt{5^2}$$
$$= -4\sqrt{49} + \sqrt{25}$$
$$= -28 + 5$$
$$= -23$$

$$(\sqrt{12} + \sqrt{12})^2$$
$$= (\sqrt{12} + \sqrt{12}) \cdot (\sqrt{12} + \sqrt{12})$$
$$= \sqrt{12^2} + \sqrt{12^2} + \sqrt{12^2} + \sqrt{12^2}$$
$$= 12 + 12 + 12 + 12$$
$$= 48$$

$$(\sqrt{4 \cdot 3})^2$$
$$= \sqrt{12} \cdot \sqrt{12}$$
$$= \sqrt{144}$$
$$= 12$$

$$(\sqrt{3} - \sqrt{12})(\sqrt{3} + \sqrt{12})$$
$$= \sqrt{3^2} + \sqrt{36} - \sqrt{36} - \sqrt{12^2}$$
$$= \sqrt{9} - \sqrt{144}$$
$$= 3 - 12$$
$$= -9$$

Simplify Radical Expressions

$$(-2\sqrt{7} + \sqrt{5})(2\sqrt{7} + \sqrt{5})$$

$$(\sqrt{12} + \sqrt{12})^2$$

$$(\sqrt{4 \cdot 3})^2$$

$$(\sqrt{3} - \sqrt{12})(\sqrt{3} + \sqrt{12})$$

Simplify Radical Expressions

$$4(\sqrt{2} - 3)^2$$
$$= 4 \cdot (\sqrt{2} - 3) \cdot (\sqrt{2} - 3)$$
$$= 4 \cdot (\sqrt{4} - 3\sqrt{2} - 3\sqrt{2} + 9)$$
$$= 4 \cdot (2 - 6\sqrt{2} + 9)$$
$$= 8 - 24\sqrt{2} + 36)$$
$$= 44 - 24\sqrt{2}$$

$$3(\sqrt{3} + 5)^2$$
$$= 3 \cdot (\sqrt{3} + 5) \cdot (\sqrt{3} + 5)$$
$$= 3 \cdot (\sqrt{9} + 5\sqrt{3} + 5\sqrt{3} + 25)$$
$$= 3 \cdot (3 + 10\sqrt{3} + 25)$$
$$= 9 + 30\sqrt{3} + 75$$
$$= 84 + 30\sqrt{3}$$

$$5(\sqrt{9} + 2)^2$$
$$= 5 \cdot (\sqrt{9} + 2) \cdot (\sqrt{9} + 2)$$
$$= 5 \cdot (\sqrt{81} + 2\sqrt{9} + 2\sqrt{9} + 4)$$
$$= 5 \cdot (9 + 6 + 6 + 4)$$
$$= 5 \cdot 25$$
$$= 125$$

Simplify Radical Expressions

$$4\left(\sqrt{2} - 3\right)^2$$

$$3\left(\sqrt{3} + 5\right)^2$$

$$5\left(\sqrt{9} + 2\right)^2$$

Simplify Radical Expressions

$2(\sqrt{11} - 7)^2$

$= 2 \cdot (\sqrt{11} - 7) \cdot (\sqrt{11} - 7)$

$= 2 \cdot (\sqrt{121} - 7\sqrt{11} - 7\sqrt{11} + 49)$

$= 2 \cdot (11 - 14\sqrt{11} + 49)$

$= 22 - 28\sqrt{11} + 98)$

$= 120 - 28\sqrt{11}$

$8(\sqrt{8} + 8)^2$

$= 8 \cdot (\sqrt{8} + 8) \cdot (\sqrt{8} + 8)$

$= 8 \cdot (\sqrt{64} + 8\sqrt{8} + 8\sqrt{8} + 64)$

$= 8 \cdot (8 + 16\sqrt{8} + 64)$

$= 8 \cdot (8 + (16 \cdot 2)\sqrt{2} + 64)$

$= 8 \cdot (8 + 32\sqrt{2} + 64)$

$= 64 + 256\sqrt{2} + 512$

$= 576 + 256\sqrt{2}$

$9(\sqrt{4} + 3)^2$

$= 9 \cdot (\sqrt{4} + 3) \cdot (\sqrt{4} + 3)$

$= 9 \cdot (\sqrt{16} + 3\sqrt{4} + 3\sqrt{4} + 9)$

$= 9 \cdot (4 + 6 + 6 + 9)$

$= 9 \cdot 25$

$= 225$

Simplify Radical Expressions

$$2\left(\sqrt{11} - 7\right)^2$$

$$8\left(\sqrt{8} + 8\right)^2$$

$$9\left(\sqrt{4} + 3\right)^2$$

Simplify Radical Expressions

$$3(\sqrt{2} - 2)^2$$
$$= 3 \cdot (\sqrt{2} - 2) \cdot (\sqrt{2} - 2)$$
$$= 3 \cdot (\sqrt{4} - 2\sqrt{2} - 2\sqrt{2} + 4)$$
$$= 3 \cdot (2 - 4\sqrt{2} + 4)$$
$$= 6 - 12\sqrt{2} + 12$$
$$= 18 - 12\sqrt{2}$$

$$2(\sqrt{5} - 2)^2$$
$$= 2 \cdot (\sqrt{5} - 2) \cdot (\sqrt{5} - 2)$$
$$= 2 \cdot (\sqrt{25} - 2\sqrt{5} - 2\sqrt{5} + 4)$$
$$= 2 \cdot (5 - 4\sqrt{5} + 4)$$
$$= 10 - 8\sqrt{5} + 8$$
$$= 18 - 8\sqrt{5}$$

$$2(\sqrt{4} - 4)^2$$
$$= 2(2 - 4)^2$$
$$= 2(-2)^2$$
$$= 2 \cdot 4$$
$$= 8$$

Simplify Radical Expressions

$$3(\sqrt{2} - 2)^2$$

$$2(\sqrt{5} - 2)^2$$

$$2(\sqrt{4} - 4)^2$$

Simplify Radical Expressions

$$7(\sqrt{6} + 1)^2$$
$$= 7 \cdot (\sqrt{6} + 1) \cdot (\sqrt{6} + 1)$$
$$= 7 \cdot (\sqrt{36} + 1\sqrt{6} + 1\sqrt{6} + 1)$$
$$= 7 \cdot (6 + 2\sqrt{6} + 1)$$
$$= 42 + 14\sqrt{6} + 7$$
$$= 49 + 14\sqrt{6}$$

$$9(\sqrt{1} + 3)^2$$
$$= 9 \cdot (\sqrt{1} + 3) \cdot (\sqrt{1} + 3)$$
$$= 9 \cdot (1 + 3\sqrt{1} + 3\sqrt{1} + 9)$$
$$= 9 \cdot (1 + 6\sqrt{1} + 9)$$
$$= 9 + 54\sqrt{1} + 81$$
$$= 9 + 54 + 81$$
$$= 144$$

$$5(\sqrt{5} + 3)^2$$
$$= 5 \cdot (\sqrt{5} + 3) \cdot (\sqrt{5} + 3)$$
$$= 5 \cdot (\sqrt{25} + 3\sqrt{5} + 3\sqrt{5} + 9)$$
$$= 5 \cdot (5 + 6\sqrt{5} + 9)$$
$$= 25 + 30\sqrt{5} + 45$$
$$= 70 + 30\sqrt{5}$$

Simplify Radical Expressions

$$7(\sqrt{6} + 1)^2$$

_____
_____
_____
_____
_____
_____
_____

$$9(\sqrt{1} + 3)^2$$

_____
_____
_____
_____
_____
_____
_____

$$5(\sqrt{5} + 3)^2$$

_____
_____
_____
_____
_____
_____
_____

Simplify Radical Expressions

$$\frac{4}{\sqrt{2}}$$

$$= \frac{4 \cdot \sqrt{2}}{\sqrt{2} \cdot \sqrt{2}}$$

$$= \frac{4\sqrt{2}}{\sqrt{4}}$$

$$= \frac{4\sqrt{2}}{2}$$

$$= 2\sqrt{2}$$

$$\frac{6}{\sqrt{3}}$$

$$= \frac{6 \cdot \sqrt{3}}{\sqrt{3} \cdot \sqrt{3}}$$

$$= \frac{6\sqrt{3}}{\sqrt{9}}$$

$$= \frac{6\sqrt{3}}{3}$$

$$= 2\sqrt{3}$$

$$\frac{10}{\sqrt{5}}$$

$$= \frac{10 \cdot \sqrt{5}}{\sqrt{5} \cdot \sqrt{5}}$$

$$= \frac{10\sqrt{5}}{\sqrt{25}}$$

$$= \frac{10\sqrt{5}}{5}$$

$$= 2\sqrt{5}$$

Simplify Radical Expressions

$$\frac{4}{\sqrt{2}}$$

_____

_____

_____

_____

_____

_____

_____

$$\frac{6}{\sqrt{3}}$$

_____

_____

_____

_____

_____

_____

_____

$$\frac{10}{\sqrt{5}}$$

_____

_____

_____

_____

_____

_____

_____

Simplify Radical Expressions

$$\frac{4}{\sqrt{6}}$$

$$= \frac{4 \cdot \sqrt{6}}{\sqrt{6} \cdot \sqrt{6}}$$

$$= \frac{4\sqrt{6}}{\sqrt{36}}$$

$$= \frac{4\sqrt{6}}{6}$$

$$= \frac{2\sqrt{6}}{3}$$

$$\frac{3}{\sqrt{3}}$$

$$= \frac{3 \cdot \sqrt{3}}{\sqrt{3} \cdot \sqrt{3}}$$

$$= \frac{3\sqrt{3}}{\sqrt{9}}$$

$$= \frac{3\sqrt{3}}{3}$$

$$= \sqrt{3}$$

$$\frac{7}{\sqrt{7}}$$

$$= \frac{7 \cdot \sqrt{7}}{\sqrt{7} \cdot \sqrt{7}}$$

$$= \frac{7\sqrt{7}}{\sqrt{49}}$$

$$= \frac{7\sqrt{7}}{7}$$

$$= \sqrt{7}$$

Simplify Radical Expressions

$$\frac{4}{\sqrt{6}}$$

$$\frac{3}{\sqrt{3}}$$

$$\frac{7}{\sqrt{7}}$$

Definitions

Radical	An expression that uses a root, such as square root or cube root.
Radicand	The number inside a root symbol.
Rationalize	A process in which the denominator of a fraction is rewritten to only contain rational numbers.
Rational	A number that can be expressed as a ratio of two integers.
Reciprocal	A number in which another number multiplied to will give a product of 1. The reciprocal of $\frac{2}{3}$ is $\frac{3}{2}$.
Inverse	Opposite or reverse mathematical operations. The inverse to addition is subtraction. The inverse to division is multiplication.
Exponent	A small number placed in the upper-right of a base number to show how many times the base number is multiplied by itself.

Did you know......

The average score on the math section of the SAT in 2011 was about 510 out of 800. Proof that there are lots of unsolved math problems.

Navier-Stokes equations are used to estimate turbulent fluid flows around aircraft and also in the bloodstream. But the math behind them still isn't understood.

The Riemann hypothesis claims there is a hidden pattern to the distribution of prime numbers and is regarded as the most significant unsolved problem in mathematics.

If you fold a piece of paper 103 times, it will be as thick as the universe.

If you could drive to the sun at 55 mph, it would take approximately 193 years to get there.

It should take no more than 20 moves to solve a Rubiks cube no matter which of the 43 quintillion possible starting positions you begin with.

18 is the only number that is twice the sum of its digits.

Any number to the power 0, a^0, is 1. But what is 0 to the power 0? $0^0 = ?$

Simplify Radical Expressions

$$\frac{4 + \sqrt{3}}{5 - \sqrt{3}}$$

$$= \frac{4 + \sqrt{3}}{5 - \sqrt{3}} \cdot \frac{5 + \sqrt{3}}{5 + \sqrt{3}}$$

$$= \frac{20 + 4\sqrt{3} + 5\sqrt{3} + \sqrt{3^2}}{25 + 5\sqrt{3} - 5\sqrt{3} - \sqrt{3^2}}$$

$$= \frac{20 + 4\sqrt{3} + 5\sqrt{3} + 3}{25 + 5\sqrt{3} - 5\sqrt{3} - 3} \qquad = \frac{23 + 9\sqrt{3}}{22}$$

$$\frac{4 - \sqrt{11}}{2 + \sqrt{11}}$$

$$= \frac{4 - \sqrt{11}}{2 + \sqrt{11}} \cdot \frac{2 - \sqrt{11}}{2 - \sqrt{11}}$$

$$= \frac{8 - 4\sqrt{11} - 2\sqrt{11} + \sqrt{11^2}}{4 - 2\sqrt{11} + 2\sqrt{11} - \sqrt{11^2}}$$

$$= \frac{8 - 4\sqrt{11} - 2\sqrt{11} + 11}{4 - 2\sqrt{11} + 2\sqrt{11} - 11} \qquad = \frac{-19 + 6\sqrt{11}}{7}$$

$$\frac{-3 + \sqrt{3}}{2 - \sqrt{3}}$$

$$= \frac{-3 + \sqrt{3}}{2 - \sqrt{3}} \cdot \frac{2 + \sqrt{3}}{2 + \sqrt{3}}$$

$$= \frac{-6 - 3\sqrt{3} + 2\sqrt{3} + \sqrt{3^2}}{4 + 2\sqrt{3} - 2\sqrt{3} - \sqrt{3^2}}$$

$$= \frac{-6 - 3\sqrt{3} + 2\sqrt{3} + 3}{4 + 2\sqrt{3} - 2\sqrt{3} - 3} \qquad = -3 - \sqrt{3}$$

Simplify Radical Expressions

$$\frac{4 + \sqrt{3}}{5 - \sqrt{3}}$$

$$\frac{4 - \sqrt{11}}{2 + \sqrt{11}}$$

$$\frac{-3 + \sqrt{3}}{2 - \sqrt{3}}$$

Simplify Radical Expressions

$$\frac{2 + \sqrt{5}}{3 - \sqrt{5}}$$

$$= \frac{2 + \sqrt{5}}{3 - \sqrt{5}} \cdot \frac{3 + \sqrt{5}}{3 + \sqrt{5}}$$

$$= \frac{6 + 2\sqrt{5} + 3\sqrt{5} + \sqrt{25}}{9 + 3\sqrt{5} - 3\sqrt{5} - \sqrt{25}}$$

$$= \frac{6 + 2\sqrt{5} + 3\sqrt{5} + 5}{9 + 3\sqrt{5} - 3\sqrt{5} - 5} \qquad = \frac{11 + 5\sqrt{5}}{4}$$

$$\frac{4 + \sqrt{7}}{3 + \sqrt{7}}$$

$$= \frac{4 + \sqrt{7}}{3 + \sqrt{7}} \cdot \frac{3 - \sqrt{7}}{3 - \sqrt{7}}$$

$$= \frac{12 - 4\sqrt{7} + 3\sqrt{7} - \sqrt{7^2}}{9 - 3\sqrt{7} + 3\sqrt{7} - \sqrt{7^2}}$$

$$= \frac{12 - 4\sqrt{7} + 3\sqrt{7} - 7}{9 - 3\sqrt{7} + 3\sqrt{7} - 7} \qquad = \frac{5 - \sqrt{7}}{2}$$

$$\frac{1 + \sqrt{6}}{3 - \sqrt{6}}$$

$$= \frac{1 + \sqrt{6}}{3 - \sqrt{6}} \cdot \frac{3 + \sqrt{6}}{3 + \sqrt{6}}$$

$$= \frac{3 + \sqrt{6} + 3\sqrt{6} + \sqrt{6^2}}{9 + 3\sqrt{6} - 3\sqrt{6} - \sqrt{6^2}}$$

$$= \frac{3 + \sqrt{6} + 3\sqrt{6} + 6}{9 + 3\sqrt{6} - 3\sqrt{6} - 6} \qquad = \frac{9 + 4\sqrt{6}}{3}$$

Simplify Radical Expressions

$$\frac{2 + \sqrt{5}}{3 - \sqrt{5}}$$

$$\frac{4 + \sqrt{7}}{3 + \sqrt{7}}$$

$$\frac{1 + \sqrt{6}}{3 - \sqrt{6}}$$

Simplify Radical Expressions

$$\frac{\sqrt{2} - 3\sqrt{3}}{\sqrt{3}}$$

$$= \frac{\sqrt{2} - 3\sqrt{3}}{\sqrt{3}} \cdot \frac{\sqrt{3}}{\sqrt{3}}$$

$$= \frac{\sqrt{6} - 3\sqrt{9}}{\sqrt{9}}$$

$$= \frac{\sqrt{6} - (3 \cdot 3)}{3} \qquad = \frac{\sqrt{3} - 9}{3}$$

$$\frac{2 + \sqrt{6}}{2 + \sqrt{3}}$$

$$= \frac{2 + \sqrt{6}}{2 + \sqrt{3}} \cdot \frac{2 - \sqrt{3}}{2 - \sqrt{3}}$$

$$= \frac{4 - 2\sqrt{3} + 2\sqrt{6} - \sqrt{18}}{4 - 2\sqrt{3} + 2\sqrt{3} - \sqrt{9}}$$

$$= \frac{4 - 2\sqrt{3} + 2\sqrt{6} - 3\sqrt{2}}{4 - 3\sqrt{6} - 3\sqrt{6} - 3} \qquad = 4 - 2\sqrt{3} + 2\sqrt{6} - 3\sqrt{2}$$

$$\frac{4}{\sqrt{2} - 2}$$

$$= \frac{4}{\sqrt{2} - 2} \cdot \frac{\sqrt{2} + 2}{\sqrt{2} + 2}$$

$$= \frac{4\sqrt{2} + 8}{\sqrt{2^2} + 2\sqrt{2} - 2\sqrt{2} - 4}$$

$$= \frac{4\sqrt{2} + 8}{2 + 2\sqrt{2} - 2\sqrt{2} - 4} \qquad = -2\sqrt{2} - 4$$

Simplify Radical Expressions

$$\frac{\sqrt{2} - 3\sqrt{3}}{\sqrt{3}}$$

$$\frac{2 + \sqrt{6}}{2 + \sqrt{3}}$$

$$\frac{4}{\sqrt{2} - 2}$$

Solve for x

$$3 + \sqrt{5x + 6} = 12$$
$$\sqrt{5x + 6} = 12 - 3$$
$$\sqrt{5x + 6} = 9$$

square both sides $5x + 6 = 81$
$$5x = 81 - 6$$
$$5x = 75$$
$$x = \frac{75}{5}$$
$$x = 15$$

$$4 + \sqrt{3x - 5} = 9$$
$$\sqrt{3x - 5} = 9 - 4$$
$$\sqrt{3x - 5} = 5$$

square both sides $3x - 5 = 25$
$$3x = 25 + 5$$
$$3x = 30$$
$$x = \frac{30}{3}$$
$$x = 10$$

$$7 + \sqrt{2x + 6} = 13$$
$$\sqrt{2x + 6} = 13 - 7$$
$$\sqrt{2x + 6} = 6$$

square both sides $2x + 6 = 36$
$$2x = 36 - 6$$
$$2x = 30$$
$$x = \frac{30}{2}$$
$$x = 15$$

Solve for x

$$3 + \sqrt{5x + 6} = 12$$

$$4 + \sqrt{3x - 5} = 9$$

$$7 + \sqrt{2x + 6} = 13$$

Solve for x

$$5 - \sqrt{10x - 4} = -1$$

$$-\sqrt{10x - 4} = -1 - 5$$

$$-\sqrt{10x - 4} = -6$$

square both sides $10x - 4 = 36$

$$10x = -36 + 4$$

$$10x = 40$$

$$x = \frac{40}{10}$$

$$x = 4$$

$$8 - \sqrt{7x - 7} = 1$$

$$-\sqrt{7x - 7} = 1 - 8$$

$$-\sqrt{7x - 7} = -7$$

square both sides $7x - 7 = 49$

$$7x = 49 - 7$$

$$7x = 56$$

$$x = \frac{56}{7}$$

$$x = 8$$

$$-3 + \sqrt{4x - 3} = 2$$

$$\sqrt{4x - 3} = 2 + 3$$

$$\sqrt{4x - 3} = 5$$

square both sides $4x - 3 = 25$

$$4x = 25 + 3$$

$$4x = 28$$

$$x = \frac{28}{4}$$

$$x = 7$$

Solve for x

$$5 - \sqrt{10x - 4} = -1$$

$$8 - \sqrt{7x - 7} = 1$$

$$-3 + \sqrt{4x - 3} = 2$$

Solve for x

$$12 + \sqrt{5x + 6} = 18$$

$$\sqrt{5x + 6} = 18 - 12$$

$$\sqrt{5x + 6} = 6$$

square both sides $5x + 6 = 36$

$$5x = 36 - 6$$

$$5x = 30$$

$$x = \frac{30}{5}$$

$$x = 6$$

$$14 - \sqrt{9x + 9} = 5$$

$$-\sqrt{9x + 9} = 5 - 14$$

$$-\sqrt{9x + 9} = -9$$

square both sides $9x + 9 = 81$

$$9x = 81 - 9$$

$$9x = 72$$

$$x = \frac{72}{9}$$

$$x = 8$$

$$-7 + \sqrt{20 - x} = -3$$

$$\sqrt{20 - x} = -3 + 7$$

$$\sqrt{20 - x} = 4$$

square both sides $20 - x = 16$

$$-x = 16 - 20$$

$$-x = -4$$

$$x = 4$$

Solve for x

$$12 + \sqrt{5x + 6} = 18$$

$$14 - \sqrt{9x + 9} = 5$$

$$-7 + \sqrt{20 - x} = -3$$

Solve for x

$$9 - \sqrt{28 - 2x} = 5$$

$$-\sqrt{28 - 2x} = 5 - 9$$

$$-\sqrt{28 - 2x} = -4$$

square both sides $\quad 28 - 2x = 16$

$$-2x = 16 - 28$$

$$-2x = -12$$

$$x = \frac{-12}{-2}$$

$$x = 6$$

$$11 + \sqrt{8x - 7} = 20$$

$$\sqrt{8x - 7} = 20 - 11$$

$$\sqrt{8x - 7} = 9$$

square both sides $\quad 8x - 7 = 81$

$$8x = 81 + 7$$

$$8x = 88$$

$$x = \frac{88}{8}$$

$$x = 11$$

$$\sqrt{3x - 5} - 9 = -5$$

$$\sqrt{3x - 5} = -5 + 9$$

$$\sqrt{3x - 5} = 4$$

square both sides $\quad 3x - 5 = 16$

$$3x = 16 + 5$$

$$3x = 21$$

$$x = \frac{21}{3}$$

$$x = 7$$

Solve for x

$$9 - \sqrt{28 - 2x} = 5$$

$$11 + \sqrt{8x - 7} = 20$$

$$\sqrt{3x - 5} - 9 = -5$$

Solve for x

$$1 + \sqrt{3x - 2} = 14$$

$$\sqrt{3x - 2} = 14 - 1$$

$$\sqrt{3x - 2} = 13$$

square both sides $\quad 3x - 2 = 169$

$$3x = 169 + 2$$

$$3x = 171$$

$$x = \frac{171}{3}$$

$$x = 57$$

$$\sqrt{5x - 4} - 9 = 7$$

$$\sqrt{5x - 4} = 7 + 9$$

$$\sqrt{5x - 4} = 16$$

square both sides $\quad 5x - 4 = 256$

$$5x = 256 + 4$$

$$5x = 260$$

$$x = \frac{260}{5}$$

$$x = 52$$

$$\sqrt{7x - 3} + 2 = 21$$

$$\sqrt{7x - 3} = 21 - 2$$

$$\sqrt{7x - 3} = 19$$

square both sides $\quad 7x - 3 = 361$

$$7x = 361 + 3$$

$$7x = 364$$

$$x = \frac{364}{7}$$

$$x = 52$$

Solve for x

$$1 + \sqrt{3x - 2} = 14$$

$$\sqrt{5x - 4} - 9 = 7$$

$$\sqrt{7x - 3} + 2 = 21$$

Solve for x

$$7 + \sqrt{2x + 5} = 32$$
$$\sqrt{2x + 5} = 32 - 7$$
$$\sqrt{2x + 5} = 25$$

square both sides
$$2x + 5 = 625$$
$$2x = 625 - 5$$
$$2x = 620$$
$$x = \frac{620}{2}$$
$$x = 310$$

$$\tfrac{1}{2}\sqrt{3x + 8} = 2$$
$$\sqrt{3x + 8} = 2 \div 0.5$$
$$\sqrt{3x + 8} = 4$$

square both sides
$$3x + 8 = 16$$
$$3x = 16 - 8$$
$$3x = 8$$
$$x = \frac{8}{3}$$
$$x = 2\tfrac{2}{3}$$

$$3\sqrt{5x - 3} = 3$$
$$\sqrt{5x - 3} = 3 \div 3$$
$$\sqrt{5x - 3} = 1$$

square both sides
$$5x - 3 = 1$$
$$5x = 1 + 3$$
$$5x = 4$$
$$x = \frac{4}{5}$$

Solve for x

$$7 + \sqrt{2x + 5} = 32$$

_____
_____
_____
_____
_____
_____
_____

$$\frac{1}{2}\sqrt{3x + 8} = 2$$

_____
_____
_____
_____
_____
_____

$$3\sqrt{5x - 3} = 3$$

_____
_____
_____
_____
_____
_____

Solve for x

$$\frac{1}{4}\sqrt{2x-4} = 1$$
$$\sqrt{2x-4} = 1 \div 0.25$$
$$\sqrt{2x-4} = 4$$

square both sides $2x - 4 = 16$
$$2x = 16 + 4$$
$$2x = 20$$
$$x = \frac{20}{2}$$
$$x = 10$$

$$2\sqrt{5x+9} = 6$$
$$\sqrt{5x+9} = 6 \div 2$$
$$\sqrt{5x+9} = 3$$

square both sides $5x + 9 = 9$
$$5x = 9 - 9$$
$$5x = 0$$
$$x = \frac{0}{5}$$
$$x = 0$$

$$\sqrt{x-10} + 1 = 2$$
$$\sqrt{x-10} = 2 - 1$$
$$\sqrt{x-10} = 1$$

square both sides $x - 10 = 1$
$$x = 1 + 10$$
$$x = 11$$

Solve for x

$$\frac{1}{4}\sqrt{2x - 4} = 1$$

$$2\sqrt{5x + 9} = 6$$

$$\sqrt{x - 10} + 1 = 2$$

Solve for x

$$\sqrt{12x + 9} - 3 = 18$$
$$\sqrt{12x + 9} = 18 + 3$$
$$\sqrt{12x + 9} = 21$$

square both sides $12x + 9 = 441$
$$12x = 441 - 9$$
$$12x = 432$$
$$x = \frac{432}{12}$$
$$x = 36$$

$$\sqrt{-2x + 8} - 1 = 3$$
$$\sqrt{-2x + 8} = 3 + 1$$
$$\sqrt{-2x + 8} = 4$$

square both sides $-2x + 8 = 16$
$$-2x = 16 - 8$$
$$-2x = 8$$
$$x = \frac{8}{-2}$$
$$x = -4$$

$$-\sqrt{9x - 1} + 7 = 6$$
$$-\sqrt{9x - 1} = 6 - 7$$
$$-\sqrt{9x - 1} = -1$$

square both sides $9x - 1 = 1$
$$9x = 1 + 1$$
$$9x = 2$$
$$x = \frac{2}{9}$$

Solve for x

$$\sqrt{12x + 9} - 3 = 18$$

$$\sqrt{-2x + 8} - 1 = 3$$

$$-\sqrt{9x - 1} + 7 = 6$$

Solve for x

$$-2\sqrt{4x - 2} = -2$$
$$\sqrt{4x - 2} = 2 \div -2$$
$$\sqrt{4x - 2} = -1$$

square both sides $4x - 2 = 1$

$$4x = 1 + 2$$
$$4x = 3$$
$$x = \frac{3}{4}$$

$$-5\sqrt{3x + 6} = -30$$
$$\sqrt{3x + 6} = -30 \div 5$$
$$\sqrt{3x + 6} = -6$$

square both sides $3x + 6 = 36$

$$3x = 36 - 6$$
$$3x = 30$$
$$x = \frac{30}{3}$$
$$x = 10$$

$$10\sqrt{5x - 4} = 40$$
$$\sqrt{5x - 4} = 40 \div 10$$
$$\sqrt{5x - 4} = 4$$

square both sides $5x - 4 = 16$

$$5x = 16 + 4$$
$$5x = 20$$
$$x = \frac{20}{5}$$
$$x = 4$$

Solve for x

$$-2\sqrt{4x-2} = -2$$

$$-5\sqrt{3x+6} = -30$$

$$10\sqrt{5x-4} = 40$$

Solve for x

$$\tfrac{1}{2}\sqrt{2x + 6} = 3$$

$$\sqrt{2x + 6} = 3 \div 0.5$$

$$\sqrt{2x + 6} = 6$$

square both sides $2x + 6 = 36$

$$2x = 36 - 6$$

$$2x = 30$$

$$x = \tfrac{30}{2}$$

$$x = 15$$

$$\tfrac{3}{4}\sqrt{2x - 4} = 3$$

$$\sqrt{2x - 4} = 3 \div 0.75$$

$$\sqrt{2x - 4} = 4$$

square both sides $2x - 4 = 16$

$$2x = 16 + 4$$

$$2x = 20$$

$$x = \tfrac{20}{2}$$

$$x = 10$$

$$\tfrac{3}{4}\sqrt{x + 1} = 3$$

$$\sqrt{x + 1} = 3 \div 0.75$$

$$\sqrt{x + 1} = 4$$

square both sides $x + 1 = 16$

$$x = 16 - 1$$

$$x = 15$$

Solve for x

$$\frac{1}{2}\sqrt{2x + 6} = 3$$

$$\frac{3}{4}\sqrt{2x - 4} = 3$$

$$\frac{3}{4}\sqrt{x + 1} = 3$$

Solve for x

$$\sqrt{9} + \sqrt{2x + 4} = 7$$
$$\sqrt{2x + 4} = 7 - \sqrt{9}$$
$$\sqrt{2x + 4} = 4$$

square both sides $2x + 4 = 16$

$$2x = 16 - 4$$
$$2x = 12$$
$$x = \frac{12}{2}$$
$$x = 6$$

$$\sqrt{16} + \sqrt{3x - 5} = 9$$
$$\sqrt{3x - 5} = 9 - \sqrt{16}$$
$$\sqrt{3x - 5} = 5$$

square both sides $3x - 5 = 25$

$$3x = 25 + 5$$
$$3x = 30$$
$$x = \frac{30}{3}$$
$$x = 10$$

$$\sqrt{4} + \sqrt{5x + 16} = 8$$
$$\sqrt{5x + 16} = 8 - \sqrt{4}$$
$$\sqrt{5x + 16} = 6$$

square both sides $5x + 16 = 36$

$$5x = 36 - 16$$
$$5x = 20$$
$$x = \frac{20}{5}$$
$$x = 4$$

Solve for x

$$\sqrt{9} + \sqrt{2x + 4} = 7$$

$$\sqrt{16} + \sqrt{3x - 5} = 9$$

$$\sqrt{4} + \sqrt{5x + 16} = 8$$

Solve for x

$$\sqrt{6x^2 - 2} = 2x$$

square both sides $6x^2 - 2 = 4x^2$

$$6x^2 - 4x^2 = 2$$
$$2x^2 = 2$$
$$x^2 = \frac{2}{2}$$
$$x^2 = 1$$
$$x = \sqrt{1}$$
$$x = 1$$

$$\sqrt{5x^2 - 8} = 2x$$

square both sides $5x^2 - 8 = 4x^2$

$$5x^2 - 4x^2 = 8$$
$$1x^2 = 8$$
$$x^2 = \frac{8}{1}$$
$$x = \sqrt{8}$$
$$x = \sqrt{4} \cdot \sqrt{2}$$
$$x = 2\sqrt{2}$$

$$\sqrt{9x^2 + 7} = 4x$$

square both sides $9x^2 + 7 = 16x^2$

$$9x^2 - 16x^2 = -7$$
$$-7x^2 = -7$$
$$x^2 = \frac{-7}{-7}$$
$$x^2 = 1$$
$$x = \sqrt{1}$$
$$x = 1$$

Solve for x

$$\sqrt{6x^2 - 2} = 2x$$

$$\sqrt{5x^2 - 8} = 2x$$

$$\sqrt{9x^2 + 7} = 4x$$

Solve for x

$$\sqrt{3x^2 + 21} = 2x$$

square both sides $\quad 3x^2 + 21 = 4x^2$

$$3x^2 - 4x^2 = -21$$

$$-1x^2 = -21$$

$$x^2 = \frac{-21}{-1}$$

$$x^2 = 21$$

$$x = \sqrt{21}$$

$$\sqrt{5x^2 - 1} = x$$

square both sides $\quad 5x^2 - 1 = x^2$

$$5x^2 - x^2 = 1$$

$$4x^2 = 1$$

$$x^2 = \frac{1}{4}$$

$$x = \sqrt{0.25}$$

$$x = \frac{1}{2}$$

$$\sqrt{3x^2 - 4} = x$$

square both sides $\quad 3x^2 - 4 = x^2$

$$3x^2 - x^2 = 4$$

$$2x^2 = 4$$

$$x^2 = \frac{4}{2}$$

$$x^2 = 2$$

$$x = \sqrt{2}$$

Solve for x

$$\sqrt{3x^2 + 21} = 2x$$

$$\sqrt{5x^2 - 1} = x$$

$$\sqrt{3x^2 - 4} = x$$

Solve for x

$$6 + 3x = \sqrt{2x + 12} + 2x$$

$$
\begin{aligned}
6 + 3x - 2x &= \sqrt{2x + 12} \\
6 + 1x &= \sqrt{2x + 12} \\
(x + 6)^2 &= 2x + 12 \qquad \text{\textit{square both sides}} \\
(x + 6)(x + 6) &= 2x + 12 \qquad \text{\textit{Use F.O.I.L}} \\
x^2 + 12x + 36 &= 2x + 12 \\
x^2 + 12x + 36 - 2x - 12 &= 0 \\
x^2 + 10x + 24 &= 0 \\
(x + 4)(x + 6) & \qquad \text{\textit{Factor}} \\
x = -4 \quad \text{or} \quad x &= -6
\end{aligned}
$$

Only -4 works in equation

$$2 + 4x = \sqrt{4x + 8} + 3x$$

$$
\begin{aligned}
2 + 4x - 3x &= \sqrt{4x + 8} \\
2 + 1x &= \sqrt{4x + 8} \\
(x + 2)^2 &= 4x + 8 \qquad \text{\textit{square both sides}} \\
(x + 2)(x + 2) &= 4x + 8 \qquad \text{\textit{Use F.O.I.L}} \\
x^2 + 4x + 4 &= 4x + 8 \\
x^2 + 4x + 4 - 4x - 8 &= 0 \\
x^2 - 4 &= 0 \\
(x + 2)(x - 2) & \qquad \text{\textit{Factor}} \\
x = -2 \quad \text{or} \quad x &= 2
\end{aligned}
$$

Either works in equation

Solve for x

$$6 + 3x = \sqrt{2x + 12} + 2x$$

$$2 + 4x = \sqrt{4x + 8} + 3x$$

Solve for x

$$2 + 2x = \sqrt{2x + 12} + x$$

$$
\begin{aligned}
2 + 2x - x &= \sqrt{2x + 12} \\
2 + 1x &= \sqrt{2x + 12} \\
(x + 2)^2 &= 2x + 12 \qquad \textit{square both sides} \\
(x + 2)(x + 2) &= 2x + 12 \qquad \textit{Use F.O.I.L} \\
x^2 + 4x + 4 &= 2x + 12
\end{aligned}
$$

$$x^2 + 4x + 4 - 2x - 12 = 0$$
$$x^2 + 2x - 8 = 0$$
$$(x + 4)(x - 2) \qquad \textit{Factor}$$
$$x = -4 \quad \text{or} \quad x = 2$$

Only 2 works in equation

$$1 + 2x = \sqrt{3x + 7} + x$$

$$
\begin{aligned}
1 + 2x - x &= \sqrt{3x + 7} \\
1 + 1x &= \sqrt{3x + 7} \\
(x + 1)^2 &= 3x + 7 \qquad \textit{square both sides} \\
(x + 1)(x + 1) &= 3x + 7 \qquad \textit{Use F.O.I.L} \\
x^2 + 2x + 1 &= 3x + 7
\end{aligned}
$$

$$x^2 + 2x + 1 - 3x - 7 = 0$$
$$x^2 - x - 6 = 0$$
$$(x + 2)(x - 3) \qquad \textit{Factor}$$
$$x = -2 \quad \text{or} \quad x = 3$$

Only 3 works in equation

Solve for x

$$2 + 2x = \sqrt{2x + 12} + x$$

$$1 + 2x = \sqrt{3x + 7} + x$$

Working with negative numbers

Add/Subtract

Like signs - Add

$$\begin{array}{r} 5 \\ +3 \\ \hline 8 \end{array} \qquad \begin{array}{r} -5 \\ +-3 \\ \hline -8 \end{array}$$

Unlike signs - Subtract

$$\begin{array}{r} 5 \\ -3 \\ \hline 2 \end{array} \qquad \begin{array}{r} -5 \\ +3 \\ \hline -2 \end{array}$$

Multiply/Divide

Like signs - Positive

$$\begin{array}{r} 5 \\ \times\ 3 \\ \hline 15 \end{array} \qquad \begin{array}{r} -5 \\ \times\ -3 \\ \hline 15 \end{array}$$

Unlike signs - Negative

$$\begin{array}{r} 5 \\ \times -3 \\ \hline -15 \end{array} \qquad \begin{array}{r} -5 \\ \times\ 3 \\ \hline -15 \end{array}$$

Algebra rules for arithmetic

$$a(b + c) = ab + ac$$

$$a\left(\frac{b}{c}\right) = \frac{ab}{c}$$

$$\frac{a}{b} + \frac{c}{d} = \frac{ad + bc}{bd}$$

$$\frac{a + b}{c} = \frac{a}{c} + \frac{b}{x}$$

$$\frac{ac + bc}{c} = a + b$$

Radicals

$$a^{\frac{1}{n}} = \sqrt[n]{a}$$

$$\sqrt[n]{ab} = \sqrt[n]{a} \cdot \sqrt[n]{b}$$

$$\sqrt[m]{\sqrt[n]{a}} = \sqrt[nm]{a}$$

$$\sqrt[n]{\frac{a}{b}} = \frac{\sqrt[n]{a}}{\sqrt[n]{b}}$$

About the author

Timothy Schablin is a graduate of the Hutchinson Technical College where he studied algebra, trigonometry, physics, mathematical techniques, and technical related fields. He also holds two certificates of physics from Davidson College, AP Physics I & AP Physics II: Challenging Concepts.

Timothy Schablin tutors math to 5th, 6th, 7th, and 8th graders at a local middle school. He is also a member of Minnesota MathCorps and has authored mathematical software.

Besides studying & tutoring math and physics, Timothy enjoys astronomy. He spends vacation time canoeing the Minnesota River bottom.

www.ingramcontent.com/pod-product-compliance
Lightning Source LLC
Chambersburg PA
CBHW072015230526
45468CB00021B/1566